RAND NATIONAL DEFENSE RESEARCH INSTITUTE

INCREASING FLEXIBILITY AND AGILITY AT THE NATIONAL RECONNAISSANCE OFFICE

Lessons from Modular Design, Occupational Surprise, and Commercial Research and Development Processes

Dave Baiocchi | Krista S. Langeland | D. Steven Fox

Amelia Buerkle | Jennifer Walters

Prepared for the National Reconnaissance Office

Approved for public release; distribution unlimited

The research described in this report was prepared for the National Reconnaissance Office. The research was conducted within the RAND National Defense Research Institute, a federally funded research and development center sponsored by the Office of the Secretary of Defense, the Joint Staff, the Unified Combatant Commands, the Navy, the Marine Corps, the defense agencies, and the defense Intelligence Community under Contract NRO-000-12-C-0187.

Library of Congress Control Number: 2013946197

ISBN: 978-0-8330-8102-5

The RAND Corporation is a nonprofit institution that helps improve policy and decisionmaking through research and analysis. RAND's publications do not necessarily reflect the opinions of its research clients and sponsors.

Support RAND—make a tax-deductible charitable contribution at www.rand.org/giving/contribute.html

RAND® is a registered trademark.

RAND OFFICES
SANTA MONICA, CA • WASHINGTON, DC
PITTSBURGH, PA • NEW ORLEANS, LA • JACKSON, MS • BOSTON, MA
DOHA, QA • CAMBRIDGE, UK • BRUSSELS, BE
www.rand.org

Preface

We performed this research for the Advanced Systems and Technologies (AS&T) Directorate at the National Reconnaissance Office (NRO). Today, the NRO faces an operational environment that is faster paced, more uncertain, and filled with more variables than it was even ten years ago. One of the biggest challenges now facing the Intelligence Community (IC) is that it must confront unknown threats that continue to emerge from unexpected directions.

To address these challenges, the NRO asked RAND to research ways that AS&T and the NRO could be become more responsive toward an ever-changing environment. To do this, RAND researched three different topics, each designed to address a different component within the NRO: the hardware, the people, and the organization's processes. For the hardware, we researched the benefits of modularity and developed a list of factors to help determine whether the NRO's space hardware was a good candidate for a modular architecture. For the people, we looked at how other occupations respond to unexpected events (i.e., surprise), with the goal of identifying a set of practices that could be employed by people who work in uncertain environments. To do this, we spoke with ambassadors, chief executive officers, military personnel, and health care professionals, and we report on some common methods and techniques that they use to prepare for and respond to surprise. Finally, we took a preliminary look at the organizational methods used inside other established organizations and made some observations about the motives behind their innovative processes.

The findings from this research will therefore be useful for NRO strategists as they make plans to shape their future hardware architectures, workforce, and organizational structures. The research on modularity and surprise has broader applications beyond the NRO and the IC and will therefore be useful for individuals working on hardware development and within uncertain environments, respectively.

This research was conducted within the Intelligence Policy Center of the RAND National Security Research Division (NSRD). NSRD conducts research and analysis on defense and national security topics for the U.S. and allied defense, foreign policy, homeland security, and intelligence communities and foundations and other nongovernmental organizations that support defense and national security analysis.

For more information on the Intelligence Policy Center, see http://www.rand.org/nsrd/ndri/centers/intel.html or contact the director (contact information is provided on the web page).

Comments or questions on this report should be addressed to the project leader, Dave Baiocchi, via email at baiocchi@rand.org or phone at (310) 393-0411, ext. 6658.

Contents

Summary

The U.S. Intelligence Community (IC) is now facing a larger number of unknown threats than at any other time in its history. During the Cold War, the IC faced one primary, well-identified threat, along with a few second-order concerns. Today, the biggest surprises facing the IC are likely to come from places of which the community may not even be aware.

To help address these challenges, the Advanced Systems and Technology (AS&T) Directorate at the National Reconnaissance Office (NRO) asked RAND to perform research to help it develop strategic plans that will yield insights on becoming more flexible and adaptable. We settled on three research questions, specifically designed to target three different aspects of the NRO enterprise:

- How can the NRO build more-flexible hardware?
- How can NRO personnel become better prepared to deal with uncertainty?
- How can the NRO's organizational structures be used to promote innovation and creative thought?

How Can the NRO Build More-Flexible Hardware?

To investigate this question, we first hypothesized that there are two ways to build more-flexible hardware: (1) by building in excess capability and (2) by using a modular architecture. Excess capability gives operators the freedom to develop new tactics, techniques, and procedures as needs change. However, it can be challenging to convince decisionmakers to support excess capability when budgets and resources are becoming increasingly constrained. Therefore, for this project, we investigated the suitability of implementing a modular architecture for the NRO's space systems.

Modularity Provides Flexibility, But at a Cost

Modularity is the engineering equivalent of a financial option: Like a financial option, modularity permits a product designer to invoke some flexibility in the future in exchange for a cost premium that is paid up front. For space systems, this premium is paid in the form of additional systems engineering that is needed to plan and design a set of standardized interfaces. These interfaces must be designed in their final form at the onset of the project so that the modules are ready for future use, providing the potential for added flexibility and responsiveness.

However, modular systems do not provide all this flexibility for free. Typically, a modular system will not perform each function as well as the equivalent individual (singular) systems. For example, a Swiss Army knife allows the user to carry a number of tools around in one small package. However, this flexibility comes at a price: The tools in the Swiss Army knife will never perform as well as a dedicated knife, corkscrew, or pair of scissors.

Different Classes of Systems Provide Different Levels of Functionality and Benefit

We researched several examples of modular systems and found that different classes of systems provide different levels of functionality and benefits. For example, dry-cell AA batteries and carpet are designed to be readily scaled based on user needs, but the primary functions of each never change. In contrast, an electronics breadboard with resistors, capacitors, and transistors offers nearly infinite functional possibilities to the user.

However, we observed that, while modular systems that offer changes in function are certainly more flexible, they also place greater responsibility on the user. For example, in order to use a breadboard kit to build an electronic device, the user needs a high degree of knowledge and experience. This is an important factor that designers should weigh when considering a modular architecture: The use of more-flexible systems often requires more-knowledgeable users.

NRO Space Systems Do Not Appear to Be Strong Candidates for Modularization

Our findings suggest that some systems might be better suited for modularity than others. To apply this knowledge about modular systems to the NRO, we developed a list of factors to help system designers determine if a system is a good candidate for modularity.

When we applied our factors to the NRO's space-based collection systems, we reached an inconclusive result: While some factors seem to encourage modularization, others seem to discourage it or are neutral. On one hand, the NRO faces uncertain future user needs, along with a customer base that desires a highly flexible product. Both of these factors encourage a modular architecture. On the other hand, the NRO relies on cutting-edge, state-of-the-art technologies in its systems, and these technologies do not lend themselves well to modularity. This is because rapid changes in technology can quickly outgrow the static interfaces in a modular architecture, rendering the entire system useless.

The NRO Needs to Be Able to Quantify the Value of Its Intelligence-Gathering Systems

So what can be done to move forward and make progress toward a more satisfying solution? What is really needed to provide a satisfying answer is a mathematical relationship that relates desired flexibility with the likelihood of investment gain or loss. To gain some perspective, we looked at how this calculation is done in another industry: parking garage design. The parking garage designer can easily quantify the balance between flexibility and investment risk. Revenue (in dollars) is a measure of value, and an interest rate is used to determine the change in value over time.

However, there is an important difference between commercial systems and the NRO's intelligence systems: It is very difficult to evaluate the *value* of intelligence systems and how that value *changes over time*. This observation leads to a key conclusion: It is not possible to find the optimum "knee in the curve" for implementing modularity if one is not able to assess the value of the intelligence resulting from the subject system.

How Can NRO Personnel Become Better Prepared to Deal with Uncertainty?

To investigate our second question, we started by thinking about other professionals who are regularly surrounded by uncertainty: stock traders; U.S. Navy Sea, Air, Land (SEAL) teams; and emergency room (ER) doctors. Practitioners of all three occupations must be comfortable dealing with surprise, and this idea yielded the two research questions that we sought to address in this work:

- Can people become more adept at planning for an uncertain future by studying surprise?
- Are there lessons for the IC in how different professionals respond to surprise?

To research this topic, we designed a framework to classify different professions based on the following two factors: (1) how quickly they typically have to respond to surprise and (2) the complexity of their work environment. We then conducted discussions with several professionals across a variety of fields to test our hypotheses.

We Identified Two Broad Categories of Responses to Surprise Among Different Professions

We found that most professionals who have to respond to surprises within seconds or minutes are usually skilled in touch labor—i.e., they work with their hands. This category includes surgeons, Navy SEALs, test pilots, and professional athletes. Practitioners in this category usually must control feelings of fear and anxiety when they encounter unexpected events, and they all have mental and physical rituals to help them manage these emotions.

Professionals who typically have more time to respond to surprises (e.g., hours, days, or weeks) are usually valued for their knowledge capital. This category includes chief executive officers (CEOs), ambassadors, military officers, and engineers. When encountering surprise, these practitioners must control ego, anger, and overreaction, and the most successful and agile practitioners in this category have typically developed mental rituals to help them manage these specific emotions.

The Level of Chaos in the Environment Also Affects People's Response to Surprise

We found that the level of chaos in the environment has a big effect on how people prepare for surprise. For example, those working in the most controlled environments, such as an athletic stadium, often have the luxury of being able to prepare a "what if" plan for every possible unexpected scenario because the range of possibilities is discrete and manageable. We found that professionals working in moderately chaotic environments tend to develop "what if" plans for the most likely scenarios, along with any scenario that represents an existential threat. When a professional of this sort encounters something in the environment that was not planned for, he or she relies on experience or training.

The Most Complex and Chaotic Situations Are Caused by Other Humans, Rather Than Something in the Environment

Regarding those working in the most complex environments, we arrived at an unexpected observation: All the individuals working in the most complex environments face surprises that are generated by other humans. A CEO, an ambassador, a Special Weapons and Tactics (SWAT) team captain, a Navy SEAL, and military general officers all fall within this category. We found that all of these professions face such complex operating environments—with an infinite number of things that can go wrong—that it does not make sense to develop comprehensive "what if" plans. Instead, the successful members of this group develop generalized frameworks that they can use to deal with surprise, regardless of the specifics of the surprise.

The Biggest Surprises Tend to Come from Third Parties

The final key finding from our research on surprise is that the biggest surprises are most likely to come from third parties—i.e., people and effects outside the immediate field of view. A Navy SEAL was the first to make this point to us, but nearly everyone else made the same observation.

The intuitive reason for this is that practitioners often spend a lot of time thinking about their adversaries, competitors, or key challenges and therefore develop a good understanding of how these forces are likely to behave. One way to address the threat of the unexpected third party is to conduct exercises to widen the organization's field of view and highlight potential alternative possibilities.

How Can the NRO's Organizational Processes Be Used to Promote Responsiveness and Creative Thought?

Our research on this final topic was motivated by the following objective: How did some organizations that have taken steps to become more responsive in promoting innovation and creative thought achieve this?

We looked at three companies suggested to us by NRO/AS&T: Pfizer, IBM, and Caterpillar. These companies have all been recently recognized in the media as having gone through transformations in order to better respond to pressures in the marketplace. However, each company reached a very different end state: Pfizer become more centralized, IBM started selling a completely different product, and Caterpillar became more decentralized. With all three companies looking to innovate, why did they take such different approaches?

Innovation Occurs for Many Reasons, Each Requiring a Different Approach

We found that innovation occurs for many reasons, and every situation requires a different approach. For example, one company might innovate to become more efficient (make better use of resources), another to become more effective (enhance current capabilities), and a third to become more agile (quickly adopt new technology). The reason for the innovation will help determine the approach taken.

As an example, we found that Pfizer decided to concentrate on anticancer and Alzheimer's drugs. To do this, it sold off and divested all of its unrelated properties so that it could concentrate on this high-risk, high-reward goal. In the process, it centralized its organization and processes to pursue a single mission.

By contrast, Caterpillar was interested in becoming more responsive to its customers' needs. To do this, the company decentralized and set up fully contained Caterpillar offices around the country, each containing everything needed to run the business: product experts, sales and maintenance teams, and finance and accounting personnel. In doing this, Caterpillar was able to customize its service to the local market, but this end goal required a different approach than that taken by Pfizer.

Conclusions

We conclude our research by noting that, even though all three topics appear to be very different, we observed three common lessons.

Modularity and Innovation Are Not Goals in Themselves

The first observation is that modularity and innovative methods are not goals by themselves—they are tools for meeting a particular goal. Instead of saying that the organization "needs to innovate" or "needs to implement a modular architecture," strategists should first set the priorities and the mission objectives. Then their organization will be in a position to determine what mechanisms should be used to meet the priorities.

Strategic Planning Would Be Beneficial for All Three Areas Discussed

The second observation is that success in modularity, innovation, and reacting to surprise all benefit from at least a partial ability to predict the future. Therefore, we conclude that any investments in developing strategic plans or visions, along with exercises designed to probe the future, can advance all three topics.

Solutions in All Three Areas Require Not Just Hardware, But Also People and Organizational Structures

Modularity, surprise, and innovative processes yield ways to evolve hardware, people, and corporate structures, respectively. Merely developing flexible hardware will not suffice because the hardware will require an equally flexible staff and organizational structure to design, implement, and operate it.

Acknowledgments

We are very grateful for our sponsors at the Advanced Systems and Technology Directorate within the National Reconnaissance Office. We thank Robert Brodowski and Susan Durham for their encouragement and support of this work throughout the project. We are particularly grateful to Geoffrey Torrington, who provided the project team with continuing guidance from the start.

We are also very thankful for several colleagues both inside and outside of the RAND Corporation. At the very beginning, Steve Rast contributed some key insights that eventually led us to our surprise framework. Bill Welser IV kept us on track by participating in the discussions that generated and refined our initial framework. Tim Webb and Scott Savitz provided careful reviews of an early draft, and their comments helped make this report stronger and easier to understand. Amy McGranahan worked hard at assembling the initial manuscript, and we are grateful to have such capable assistance. Finally, we appreciate the efforts of our editing team: Nora Spiering, Matt Byrd, and Steve Kistler all worked hard to ensure that this piece was properly edited, proofread, and typeset.

The observations and conclusions made within this document are solely those of the authors, as are any errors or omissions, and do not represent the official views or policies of the U.S. Intelligence Community or of the RAND Corporation.

Abbreviations

AC	alternating current
AS&T	Advanced Systems and Technology
ASW	antisubmarine warfare
CD	compact disc
CEO	chief executive officer
DoD	Department of Defense
ER	emergency room
FOV	field of view
HRO	high-reliability organization
IC	Intelligence Community
I/O	input/output
LCS	littoral combat ship
MCM	mine countermeasures
NFL	National Football League
NGO	nongovernmental organization
NPV	net present value
NRO	National Reconnaissance Office
R&D	research and development
RPD	recognition-primed decision model

SEAL   Sea, Air, Land

SLR   single-lens reflex

SUW   surface warfare

SWAT   Special Weapons and Tactics

USB   Universal Serial Bus

Investigating the Suitability of Modularity Toward National Reconnaissance Office Space Systems

---

Our Objective Was to Determine if Modularity Was Suitable for NRO Space Systems

- **We started by reviewing the academic literature and talking to subject matter experts in modularity:**
  - **What is modularity?**
  - **Why do it?**
  - **What are the ingredients of a modular architecture?**
- **We developed a list of criteria to evaluate whether systems are good candidates for modularization**
  - **We tested the list on the littoral combat ship (LCS)**
- **How do NRO space systems rate using these criteria?**
- **What questions does the NRO need to address to move forward?**

RAND

We will begin by looking at the first topic: modularity. We studied modularity because it represents one approach for creating systems that can adapt to change. This slide describes our research objective and method.

As we will show, implementing a modular architecture provides the user with a system that is more easily modified to respond to external changes, but the ability to make these changes often comes at a cost of additional up-front systems engineering.

The goal for our analysis was to take a first-order look at whether NRO systems are good candidates for modularity.

To address this question, we started by examining the academic literature and consulting with subject matter experts on modularity. We then used our findings to form a clear definition of what modularity is and how it can make a system more easily adaptable to change. After researching how systems can adapt to change using modularity, we formed an outline of the elements that are necessary for a successful modular architecture.

Next, we developed a list of first-order factors for evaluating whether or not a system is a good candidate for modularization. To test our factors, we applied them to the missions addressed by the littoral combat ship (LCS), which is widely recognized as being a modular system (Alkire et al., 2007, p. 6; O'Rourke, 2012, p. 1). We then applied our factors to NRO space systems to determine if they might benefit from a modular architecture.

This section concludes with recommended actions that the NRO can take to further determine how to incorporate modularity into its systems.

plicated systems have more parts and interdependencies, which may make them less reliable.

Interfaces. The interfaces need to be the most accommodating part of the system: As modules come and go, the interface remains embedded within the system. Therefore, the most successful interfaces will be compatible with changing technology, even when future trajectories of the technology are not yet known. This is a significant challenge when modularizing a system, but it is not an impossible feat: There are many examples of successful interfaces that have demonstrated long service lifetimes.

One example is the physical interface of the Universal Serial Bus (USB) plug, which has remained consistent for over 20 years. Despite the rapid change of technology, USB ports still use the same physical plugs, are backwards compatible, and demonstrate the possibility of anticipating future needs. In an industry known for changing every 18 months, the fact that USB remains the primary input/output (I/O) interface for consumer PCs is a testament to the original architects' ability to engineer a very flexible interface.

A second example of a successful interface is the physical plugs associated with electrical alternating current (AC) appliances. While each country does have a different standard, each of these plugs has been very robust: Appliance designers have relied on the same physical interface with the U.S. (consumer) electrical grid since the early part of the 20th century. This interface is now just as adept at handling a modern computer as it was with a 1950s-era toaster.

Test infrastructure. A test infrastructure is the third component of a successful modular system. The test infrastructure is a way to ensure that new modules will work as expected when they are inserted into the production or operational system. Developing a test infrastructure usually means building some level of a stand-alone redundant system, recognizing that this will incur additional cost. The alternative to this is to plug modules into a final production system, but this defeats the purpose of having a modular architecture because the production system will serve as a piece of test equipment. If the new module does not work, the fielded system will then have to function as a testbed, taking the system away from its regular duties. As one modularity expert with whom we spoke noted, if the method for testing new modules is just plugging them into the final system, there is no advantage to building a modular system (Baldwin, 2012).

These three components highlight an important philosophical approach that is inherent with modular system design: Every modular architecture must "freeze" some components in order to "free" others (Baldwin, 2012). This idea is evident in all aspects of modular design: Something has to be held constant in order to let other components come and go. This is a powerful idea because it suggests that the options that come with modular systems do not come without cost. In freezing components (usually the interfaces), the architect is putting constraints in place, but these constraints are (ironically) a requirement to obtain the freedom that he or she is seeking.

All of the background presented so far has suggested that modular architectures have both strengths and weaknesses. This implies that there are some missions or needs that may be more suited to modularity than others, and we set out to develop some factors that could be used to evaluate whether an arbitrary system is a good candidate for *modularity of use*. The factors that we propose in this list are a product of our research and thinking on modular theory, but the idea of developing a list of factors was motivated by Schilling (2000) and Gershenson et al. (2003).

The first factor is associated with how easy it is to physically separate the functions that will later be assigned to different modules. We use the word *component* to describe these functions, and in using this word we refer to the parts that make up the final device as seen from the end user's perspective.[2] If the system can be easily decomposed and reassembled via some components that roughly map to different functions, the system (or the mission set) is a good candidate for modularity. Conversely, the need for highly integrated parts makes modularization extremely challenging. This is because highly integrated parts will impose a high cost during the early (and subsequent) systems engineering processes, as engineers deconflict all of the integrated components to determine how all of them interact with one another. If this initial cost of added systems engineering is too high, it may overshadow any gains in flexibility that the modular system will be able to provide further down the road.

The second factor speaks to the needs of the user: Does the user prefer flexibility or highly optimized functionality? As we mentioned earlier, modular systems have the potential to provide great flexibility, with the trade-off being that modular systems will never perform as well as singular tools designed for the specific tasks.

The third factor has to do with how mature the technology is. Widely used, mature, or universal technologies are likely to have existing standards in place, and this makes modularizing these technologies less challenging. (This practice of leveraging universal technologies is sometimes referred to as *standards-based innovation*.) By contrast, emerging technologies will have less-mature standards, and it will be more challenging to design a set of enduring interfaces. For this reason, emerging and specialized technology is more difficult to modularize.

The fourth factor is related to the third: If the technology is changing in a predictable, incremental way, modularizing will allow new technology to be easily adapted (assuming that the interfaces have been designed to accommodate these changes.) However, if the trajectory is unknown or revolutionary changes are expected, modularizing could actually inhibit innovation by tying the user to a set interface that will potentially not be able to adapt to dramatic changes in technology. This factor motivates an important observation: Disruptive products are unlikely to arise out of modu-

---

[2]   Of course, the end user is likely to have very little knowledge about these components, but that is unlikely to affect the item's modularity of use. For example, the people who work in train yards coupling and uncoupling cars only need to understand how the cars' interfaces work; they do not need to understand the details of the components, such as the wheel assemblies, passenger compartments, or electrical systems.

There Are Three Choices When Deciding How to Size a Parking Garage to Meet Demand

A

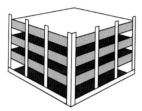

B

C

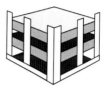

✓ **High capacity to meet future demand**
✓ **Risk of unused spaces**

✓ **Lower capacity that meets current demand**
✓ **No room for growth**

✓ **Low initial capacity meets current demand**
✓ **Spend a little more for the _option_ to add on later**

RAND

We chose the parking garage example because it offers a simple analog to the same problem that the IC is trying to solve. Specifically: Is there a way to determine how much more should be spent up front (if anything) buying options for increased flexibility in the future? As we will show, parking garage designers rely on a simple net present value (NPV) calculation to quantify the balance between investment risk and potential benefits (de Neufville and Scholtes, 2011). After we describe how parking garage designers quantify this balance, we will draw parallels to the NRO's space-based collection systems and identify what quantities are needed to repeat this approach for these systems. However, we will begin by describing the different choices that a parking garage developer has when trying to size a new garage to meet expected demand.

Assuming a simple scenario, the parking garage developer has three choices when determining what kind of garage to build. The developer could build a garage that is larger than what she or he needs today to meet current demand (option A in the slide). This choice represents the largest up-front investment, but it also provides the most excess capacity (and therefore additional revenue) should the demand ever increase. The opposite of option A is to build a small garage that exactly meets the current demand (option B). This option is the least expensive, but it provides no room for growth. If the demand for parking spaces increases, the developer will have to build a new garage. The final option is to build a garage is that sized for today's demand but also has a stronger foundation that would allow the developer to add additional floors should the demand increase in the future (option C).

The Most-Flexible Choice Is to Pay For the Option to Build More Levels Later

A

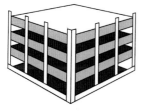

✓ **High capacity to meet future demand**
✓ **Risk of unused spaces**

B

✓ **Lower capacity that meets current demand**
✓ **No room for growth**

C

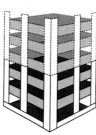

✓ **Low initial capacity meets current demand**
✓ **Spend a little more for the _option_ to add on later**

RAND

The most-flexible choice is to pay for the option to build more levels later. Option C will be less expensive than A and more expensive than B, but it allows the garage developer to spend more today for the option to add extra floors later to meet future demand. So how does the garage developer decide exactly what dimensions he or she needs, based on his or her appetite associated with the risk of investment loss (along with the desire for potential gain)? Or, to pose a question that is often asked when designing modular systems: Where is the "knee" in the modularity curve? How many options should the developer purchase to achieve a suitable level of future flexibility?

The Parking Garage Designer Can Quantify the Balance Between Flexibility and Investment Risk

- **What can we learn from the method used to design parking garages?**

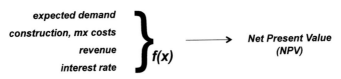

- **Result: relates likelihood of gaining or losing $N for a desired level of flexibility.**
- **When compared to a rigid design, a flexible garage offers the ability to better match future demand with less likelihood of taking a loss.**

RAND

So how does the parking garage developer decide how to proceed? The answer is that the developer performs an NPV calculation that allows him or her to quantify the balance between future flexibility and likelihood of investment gain or loss (de Neufville and Scholtes, 2011).

There are four key quantities that the developer uses to perform this calculation:

- The first quantity is the expected future demand; this is represented by a probability distribution.
- The developer also considers the construction and maintenance costs across the life of the garage.
- The developer determines how much to charge each car that enters the garage, and this is used to calculate revenue, given a particular demand curve.
- Finally, the developer uses an interest rate to normalize price across the lifetime of the project and determine the NPV.

If the developer chooses a single-input demand probability, he or she can calculate an NPV across the lifetime of the project for options A, B, and C. To make this calculation more useful, the developer could perform a Monte Carlo calculation, varying the demand curve each time. The result of this calculation will provide the developer

user when it makes sense to modularize or not, or to what degree a system should be modularized.

We Recommend Two Approaches to Make Further Progress on This Topic

- **Our criteria suggest inconclusive results on whether to pursue modularization.**
- **Moving forward, there are two options:**
  - **To properly determine how much flexibility (modularity) is needed, a system architect needs to be able to quantify the value of intelligence.**
  - **In the meantime, a more qualitative (although less satisfying) approach would be to identify 2 to 4 performance needs that *really* matter:**
    - **Is it possible to build modules that support the performance objectives?**
    - **What interfaces will be needed?**

RAND

We conclude this section by noting our key finding: NRO space systems are not obvious candidates for modularization. So what can be done to proceed?

The parking garage analogy yielded one suggestion for making progress: Develop methods to assess the value of space-based intelligence-gathering systems. This is the best way to progress toward developing a quantitative approach to answering the question that most system architects ask when considering a modular design: Where is the knee in the modularity curve? Or, how much modularity provides the most flexibility for a modest up-front investment? As our analogy to the parking garage demonstrated, getting a quantitative answer to these questions means solving the problem of quantifying the value of the intelligence provided by the system.

We recognize that quantifying the value of intelligence sensors is one of the IC's biggest challenges. NRO/AS&T is currently working on this problem for its own sensors, but it recognizes that more work is needed before an effective solution is developed.

In the meantime, we recommend a secondary approach for moving forward. As a starting point, the modularity experts with whom we spoke recommended thinking about what could be done if starting from scratch, without any restrictions (Baldwin, 2012; Schilling, 2012). What would the NRO design to do today's mission, if it were allowed to ignore legacy missions or capabilities that are being carried forward as forced requirements? What key performance metrics are captured by this "from scratch" system?

After identifying the appropriate mission needs, the NRO could then consider the benefits of modularity by determining if these needs would be easily modularized using the factors that we presented earlier. Does it make sense to build modules to support these needs? Is it even possible, given the existing technology trajectories? What interfaces will be needed?

We recognize that the results from this approach will never be as satisfying as the results from solving the IC's analog to the parking garage problem because it will not yield a quantitative result. Instead, it offers an interim solution for making progress before the larger challenge of addressing intelligence value is addressed. However, even though this approach offers an easier path (in return for less-satisfying answers), there are still several challenges that must be addressed. Specifically, identifying the right components to modularize to accommodate change will be a significant challenge, and modularizing when there are many unknowns can be precarious. There is a significant amount of risk in committing to any specific trajectory or performance metric when the direction of the technology is uncertain, and for this reason modularizing too early in the development of a technology is often not recommended. If the technology trajectory is misidentified, the user could be left with an obsolete system, having wasted effort on modularizing and subsequently finding it to be incompatible with newly evolved technology.

An example from the music industry, which was brought to our attention by one of the modularity experts with whom we spoke, is helpful for illustrating this point (Schilling, 2012). Up until the mid-1990s, the technologies used for music playback all progressed with increases in playback fidelity, starting with vinyl records before moving to cassette tapes and then compact discs (CDs). In the '90s, the music industry started looking for a playback medium that was of even higher fidelity than the compact disc, and they started working on a standard called Super CD. Yet, Super CDs do not exist today. What happened?

The reason that Super CDs have not been commercialized is because the music industry got it wrong: They thought that consumers wanted higher-fidelity sound, but consumers really wanted portability. While the industry was trying to settle on the standards for the Super CD, the rise of inexpensive personal computing gave rise to MP3, an electronic file format. What is interesting about this example is that the progression from records to cassettes to CDs aligns with increasing improvements in both fidelity *and* portability, but the music industry either failed to identify this parallelism or simply followed the wrong metric. The lesson is that predicting future needs is not straightforward and is highly uncertain. Choosing the wrong parameters and technology trajectories can end up inhibiting innovation and enhancement of a system and, in the worst case, can leave the user with a system with functions she or he does not need and lacking the functions that are needed.

Before we describe our method and findings in detail, it is worth noting that this research topic proved to be so rich that we have also published a stand-alone document that more completely describes this research (Baiocchi and Fox, forthcoming). For the purpose of this report, we will outline our initial hypotheses, research method, and findings. We will conclude by making observations on how our findings can be used by NRO/AS&T. Our longer document greatly expands on these topics, including additional context on our hypotheses and data collection methods, along with additional anecdotes that we gathered from all of our practitioners. In addition, we expanded the context of our findings and recommendations so they could be accessible to an audience outside the U.S. IC.

Finally, at the request of our sponsor, we purposely did not include occupations from the IC in our assessment. We did this for two reasons. First, we did not want to bias our result by including our immediate audience in our research. If we had, we may have favored conclusions toward one group or another, possibly favoring one conclusion over another. Second, the sponsor was interested in presenting the results from this research to the IC at large, and we suspect that different parts of the IC will fall into different portions of our framework. Leaving the IC out of our initial dataset means that each group can make its own assessment on where it stands within our framework and conclusions.

At the most basic level, preparing for and responding to surprises is about effective decisionmaking. Therefore, our initial approach was to begin by reviewing the pertinent decision-science literature that relates to unexpected events.

There are a number of books that address aspects of the surprise problem. Weick and Sutcliffe's *Managing the Unexpected: Resilient Performance in an Age of Uncertainty* (2007) presents a set of rules that high-reliability organizations (HROs) should follow in order to effectively mitigate the negative effects associated with unexpected events.[1] Bazerman and Watkins's *Predictable Surprises: The Disasters You Should Have Seen Coming, and How to Prevent Them* (2008), focuses on the nature of the surprise itself, and they outline a set of characteristics that are useful for identifying "predictable surprises" before they cascade into more-damaging effects. In *Thinking in Time: The Uses of History for Decision-Makers* (1986), Neustadt and May focus on the utility of using historical analogies to inform present-day decisionmaking. McCall, Lombardo, and Morrison's *Lessons of Experience: How Successful Executives Develop on the Job* (1998) discusses how important experience is for effectively responding to unexpected events.

As we would later confirm, all of the components highlighted in these works—strategies for developing effective response techniques, methods for identifying surprise indicators, development of personal and institutional experience, and the ability to

---

[1]   The term *high-reliability organizations* (HROs) is used by Weick and Sutcliffe to describe organizations that have "no choice but to function reliably. If reliability is compromised, severe harm results" (Weick and Sutcliffe, 2007). Specific examples of HROs include nuclear power plants, air traffic control systems, and hostage negotiation teams.

leverage historical analogies—are important elements of an effective preparation and response plan. However, as the starting point of our own research, we needed to understand the specifics of the decisionmaking process that professionals use when confronting surprise. This required identifying a relevant model as a starting point from which to develop testable hypotheses to probe through our discussions with expert practitioners.

In the end, we settled on Gary Klein's *Sources of Power: How People Make Decisions* (1998). Through years of field observations watching firefighters, aircraft carrier operations, and intensive care units, Klein has refined what he calls the recognition-primed decision model (RPD). The RPD provides a notional framework for how high-stakes decisions are made (Klein, 1998, p. 7).[2] As Klein notes in his exposition of the model, it fuses two processes: how decisionmakers use prior experience to recognize a situation and how decisionmakers decide on an appropriate course of action (Klein, 1998, p. 24).

Klein's model outlines three potential paths that decisionmakers follow when confronting a decision:

- The first is for executing prescribed responses to easily identified problems: The decisionmaker recognizes the change as something that has been seen before and proceeds using a previously prescribed course of action.
- The second variant is for when the problem cannot be immediately identified. In this case, the practitioner will attempt to diagnose the situation by matching observed features with past experience. If this is not immediately successful, the practitioner continues to observe the scenario to collect more data until he or she is able to determine causality.
- The third RPD variant applies when the practitioner can identify the problem but does not have a prescribed solution at hand. In this case, the decisionmaker often uses mental simulation to evaluate each potential response to identify the best option. Once the optimal solution is identified, the decisionmaker can implement that as the proper course of action.

Klein provides a helpful analogy for understanding these paths, likening decisionmaking to an "if-then" process. The first variant is when the decisionmaker knows both the "if" and the "then." The second is when the decisionmaker only knows the "then," and the third is when he or she only knows the "if" (Klein, 1998).

Klein's model therefore provides an initial framework that describes how professionals make decisions in uncertain environments. The model suggests three key aspects regarding how people respond to surprise:

---

2   Klein implicitly defines high-stakes decisions as decisions in which lives or significant resources are at stake (Klein 1998, p. 4).

- Whether conscious or subconscious, decisionmakers rely on a notional decision loop to evaluate and make decisions.
- Decisionmakers will use different mental mechanisms to respond, depending on whether the surprise is immediately recognized.
- When confronted with a situation for which they lack a prescribed solution, decisionmakers use mental simulation to test out potential responses before deciding on a course of action.

We were also interested in probing deeper into the nature of the surprise, so we drew inspiration from nine factors that Klein uses to define a naturalistic decisionmaking setting (2008). These factors are experienced decisionmakers, high-stakes scenarios, dynamic conditions, inadequate information, time pressure, ill-defined goals, poorly defined procedures, team coordination, and cue learning. Because our research objective required distinguishing between various surprise operating environments, we focused on three key factors of naturalistic decisionmaking settings: time pressure, inadequate information, and dynamic conditions. Those three factors appeared (to us) most likely to influence the approach that experienced professionals use when responding to unexpected events.

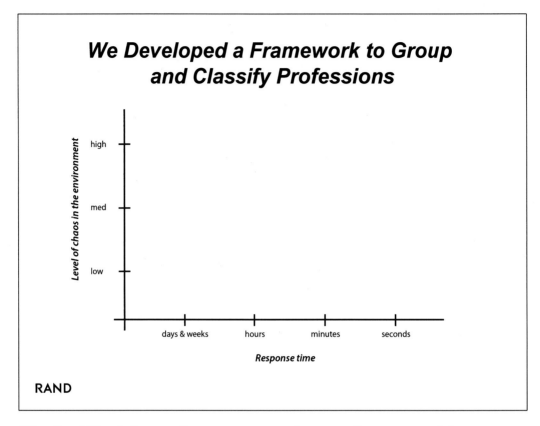

We Developed a Framework to Group and Classify Professions

RAND

We refined Klein's factors of time pressure, inadequate information, and dynamic conditions into a two-dimensional classifying framework, or "surprise space." The framework that we developed is shown above. The two axes are categorized by *typical response time* and *level of environmental chaos* (which combines the concept of a dynamic environment with one of inadequate information).

We used *typical response time* to characterize the horizontal axis, and we have four discrete points (seconds, minutes, hours, and days and weeks) that characterize an occupation's most common operating mode.[3] An example of someone who operates with a typical response time of a few seconds is a combat airplane test pilot; an example of someone who operates with a response time of days and weeks is a CEO.

We used *level of chaos in the environment* to characterize the vertical axis, and we chose three levels: low, medium, and high. We define *chaos* as a subjective measure of the frequency, diversity, and predictable orderliness of events, and this meaning seemed to best fit the quality that we were trying to assess. This definition followed from the recognition that some professions work in more controlled environments than others, and we hypothesized that this would be an important factor when considering how

---

[3]   We characterize *response time* as the span between the surprise being detected and the point at which a decision must be made to mitigate or prevent the event from becoming a more-challenging scenario.

We were fortunate in that everyone with whom we attempted to speak agreed to participate in the project. The final list included 13 representative occupations: CEO, retired U.S. ambassador, test pilot, foreign service officer, cardiothoracic surgeon, recently retired Navy SEAL, recently retired NFL coach, professional improv actor, public works/civil engineer, space mission planner/operator, recently retired Air Force lieutenant general, SWAT team commander, and emergency room (ER) physician.

We have refrained from disclosing the names of those with whom we spoke because part of our research approach was to promise our participants anonymity. Conversations lasted from 45 to 90 minutes, with approximately half conducted by phone and the other half in person. Two researchers participated in each discussion, with one taking the lead role, and the other primarily taking notes.

We had three goals for these discussions. First, we wanted to confirm that we had correctly located each profession within our axes. To do this, we asked each representative how quickly he or she had to react when responding to surprise. We also asked them questions about their operating environments to make sure that our estimate about their level of chaos was accurate. The positions shown on the previous slides reflect the final, representative locations of all the exemplar occupations.

The second goal of the discussions was to learn more about the techniques and tools each person used when responding to surprise. To do this, we asked the participants questions about how they typically responded to surprise and what (if any) professional or organizational protocols they relied on when responding. Our questions were based on our hypotheses and research objectives, but the conversations were allowed to progress organically, and they did not follow a highly structured protocol.

The final goal of these discussions was to obtain anecdotes that could be used to support and illustrate our findings. Participants offered a surprisingly rich and varied set of such stories, some of which we share in the following pages.

We conducted semistructured interviews with practitioners working in each of these professions to achieve the three goals mentioned above. Before beginning the interviews, we developed a basic protocol that outlined the key topics that we hoped to discuss in every conversation. Specifically, we focused on the following (exemplar) questions:

- What is the level of chaos and usual available response time in the typical operating environment?
- What are some other parameters and constraints of the typical operating environment?
- What are the criteria for operational success or failure?
- What role does surprise play in the operating environment?
- What does the response process look like when surprises occur?
- What resources, tools, and strategies are available for dealing with surprise, and how are they typically applied?

- How much do flexibility and finesse apply when dealing with surprise?
- To what extent are surprises viewed as opportunities rather than obstacles?
- What are the key elements needed to successfully deal with surprise? What are some examples of when those approaches failed?
- What are some characteristics of a professional in the field that allow the person to best handle surprise?
- Finally, we always gave participants the opportunity to express additional thoughts on topics that had not been covered earlier.

Data from each practitioner conversation was analyzed using qualitative methods. Both investigators discussed the content and debriefed each other within 24 hours following each conversation. We compared information provided by the professional against the relevant components of our initial conceptual model, looking for confirmation, disconfirmation, and novel insights or features. We used the descriptions that participants described about their methods and processes to test our hypotheses.

We also compared their information against that provided by other professionals, looking for subjective similarities and differences in their approaches to surprise. The goal of this particular exercise was to identify key themes across the different practitioners. Where differences in approach could be identified, we also attempted to characterize any associated differences in the surprise environment between those professions. These iterative updates to our initial hypotheses, including any tentative conclusions regarding the reasons for differences and similarities in approaches between professions, were then subsequently tested when discussing related topics with other professionals. Following the completion of our discussions with all recruited professionals, the results of this analysis were then organized and grouped by initial hypothesis.

It was not our intent to evaluate our interview data using quantitative or statistical means. For example, we did not set out to make statements like, "75 percent of all CEOs rely on the same method to respond to an unexpected event." This is because such an approach was not in alignment with our project's goal of simply identifying general trends across the professions. A second reason is that we interviewed 15 professionals in total, which is not a statistically large sample.

How did we know that interviewing 15 professionals was sufficient in meeting our goal for identifying general trends? For guidance on this, we followed a suggestion by Robert S. Weiss in *Learning from Strangers: The Art and Method of Qualitative Interview Studies* (1994): "[Y]ou stop when you encounter diminishing returns, when the information you obtain is redundant or peripheral." About two-thirds of the way through the interview process, we began to recognize familiar narratives, a good indication that our small data set was nevertheless providing empirically reliable data. This suggested that we had reached what survey managers call the "saturation point" and was a good indication that the number of interviews was sufficient for meeting our research objective.

> # *We Found That Some Response Techniques Were Common to All Practitioners*
>
> - **We observed that all successful surprise practitioners:**
>   - **Rely heavily on past experience, including experience gained via repetitive training.**
>   - **Seek to reduce the level of chaos within the operating environment**
>   - **Match the response level to the actual need**
>   - **Consistently mention that teamwork is an essential component**
>
> **RAND**

We found a number of coping strategies that were common across all of the professions.

First, we observed that all practitioners rely on experience. Experience is one of the best insurance policies against the negative effects of surprise because it provides the context necessary to recognize and respond to unexpected events. Experienced practitioners are able to quickly identify surprises and generate potential reactions and solutions based on what they have done in the past.

We also observed that all professions try to reduce the level of chaos in the operating environment, since reducing chaos also reduces the complexity and size of the solution space. Reducing chaos can be done in different ways, either by controlling the environment or spending a few extra moments (or days, depending on the profession) gathering more data on the situation.

For example, test pilots only execute one maneuver at a time. In addition, the test plane is outfitted with a suite of environmental sensors that are constantly being monitored by a room of test engineers. When the engineers detect a problem, they are able to provide the pilot with data and context that help him to focus his response procedures toward the root cause of the problem.

Similarly, the cardiothoracic surgeon with whom we spoke prefers to work with the same team of nurses and doctors for all of his procedures. In addition, he organizes his surgical tools using the same layout for every surgery, and he tends to repeatedly use pieces of monitoring equipment with which he is most familiar. Both of these examples demonstrate ways that practitioners reduce the size of the surprise space to minimize the number of things that can go wrong.

We also learned that it is best to react to surprises with a measured response to preserve further options as the surprise unfolds. As examples of this, both the CEO and the Navy SEAL used essentially the same language to tell us that it is important not to overreact when confronted by a surprise. Along with nearly all of our participants, they emphasized the importance of matching the remedy to the actual need.

Finally, we observed that teamwork plays an essential role when responding to unexpected events, even for those professions that are usually perceived as relying on individual actors, like heart surgeons or test pilots. All of the professions emphasized that they all relied heavily on teamwork to successfully deal with surprise events.

We Observed That the Level of Environmental Chaos Determines How Practitioners Prepare

- **In the most contrived environments, practitioners develop and rely on specific "what if" plans**
- **In moderately complex environments, practitioners**
  - **Prepare "what if" plans for likely and existential threats**
  - **Rely on protocols and training for all else**
- **In complex environments, practitioners**
  - **Develop robust and generalized response frameworks (and exercise them)**

RAND

We found that the level of environmental chaos strongly affects how practitioners prepare for and respond to surprise. Specifically, we found that practitioners who work in the most structured (low-chaos) environments, like athletic fields or theatrical stages in which most environmental factors are controlled, face only a finite range of possible outcomes. Because of this, these practitioners are able to plan reactions for nearly any event, and practitioners in the NFL, for example, regularly do so. For many working in low-chaos environments, it makes sense to develop comprehensive "what if" plans because the size of the surprise space is small (compared to other professions) and finite.

Practitioners who work in moderately chaotic environments like operating rooms or test plane cockpits rely on preplanned protocols for the most likely events, along with anything that represents an existential threat. As an example, the ER physician with whom we spoke indicated that he had protocols to deal with some of the most common injuries that arrive in the emergency department. However, he also recognized that there are too many unforeseen events in moderately chaotic environments to plan against every possibility, so he relied on basic response frameworks when the surprise event was not covered by a specific protocol.

In addition to preparing for the most common threats, we found that practitioners in moderately complex environments also make "what if" plans for existential threats. For example, the Mars rovers take most of their commands directly from mission managers on Earth each day. However, a small number of commands are embedded (preprogrammed) on the spacecraft to guard against existential threats. An example of a mission-ending threat would be if the rover's communications dish were ever positioned below the horizon line. If this occurred, the rover would not be able to receive commands from earth, and this would effectively represent the end of the rover's life. To address this possibility, the rovers are preprogrammed with instructions to automatically reposition the antenna back toward Earth.

The most challenging circumstances are faced by those practitioners working in highly chaotic environments, such as in a foreign embassy or behind enemy lines. Their environments are so complex and unpredictable that it does not make sense to do much planning against specific surprise events. (This is because workers in the field must deal with third parties, unpredictable weather, and unfamiliar terrain.) Instead, practitioners working in highly chaotic environments develop and exercise a general-purpose framework that can be deployed whenever a major surprise is encountered.

As an example, the former ambassador with whom we spoke referred to his framework as a "task force" and noted that the task force is the standard tool that he used whenever he encountered a surprise. Therefore, he and his staff had worked out the details of assembling a task force ahead of time. While they did not know what the specific surprise was going to be, they knew it would require office space, lines of communication, and the support of some key people inside and outside the embassy. They then prepared and practiced a process for quickly deploying this infrastructure when needed.

We Observed That Strategists Use Different Response Mechanisms Than Tacticians

- **When a surprise happens to a tactician, they are often trying to control fear and anxiety. Their response tends to involve the following steps:**
  - **Control panic**
  - **Buy time**
  - **Revert to fundamentals learned in training, allowing effective response using minimal analysis.**
- **By contrast, strategists often have to control ego, anger, and overreaction. To do this, effective strategists will**
  - **Control emotions to maintain objectivity**
  - **Take initial enabling actions**
  - **Assemble the staff**
  - **Socialize the longer-term plan**

RAND

We found that practitioners within our *tactical professions* rely on different tools than those within our *strategic professions*.[8]

Based on the data that we collected during our interviews, we found that when tacticians are surprised, they often have to deal with emotions of fear and anxiety. As a result, we found that their response loop is designed to combat those emotions. We also note that tacticians typically rely on protocols to overcome these emotions in order to respond effectively using minimal analysis. As an example, the test pilot noted that he was trained to "wind the watch" in the moments after a surprise, which refers to the menial task of winding the stem on his mechanical wristwatch. The goal of this task is to focus on a basic manual chore, with the hope that it would provide calmness and clarity in the immediate moments after a surprise occurs. Likewise, the former Navy SEAL noted that when SEAL teams are fired upon, they quickly take steps designed to control panic and buy time: They fall into rehearsed positions on the ground that help them establish a strong defensive position. In taking these positions, the team is

---

[8]   Again, we use the words *tactical* and *strategic* with specialized meanings in this document: The tactical professions are those that have to typically respond within minutes or seconds. Members of the strategic professions typically have days, weeks, or months to respond. As we stated earlier, we recognize that those individuals working in more tactical occupations certainly also engage in long-term (strategic) planning, and vice versa for the strategic professions.

able to gain confidence by quickly executing a familiar task that they have performed hundreds of times before.

In contrast to the tacticians, we found that surprising a strategist can cause anger and impulse desires to overreact. To combat this, we found that many successful strategists employ the same four-step process to counter these issues: take steps to control emotion and ego, initiate first-order enabling actions, assemble a trusted inner circle of advisors and direct reports, and disseminate a coherent longer-term response throughout the organization.

The CEO gave us the most intuitive example of a way to control emotion and ego; he noted that instead of responding immediately when surprised by an email, he places the message in his drafts folder and revisits it 24 hours later to confirm that the message is objective and constructive.

The retired three-star general officer gave us the best example of taking initial enabling actions after a surprise occurs. The general with whom we spoke was in command of some key logistics resources on September 11, 2001. As the day unfolded, no one was sure what was going on, but the general was confident that he would be called upon to provide his resources to the national command authority. Because of this, he started mobilizing a measured number of resources early on to ensure that a basic level of service would be available. This proved to be an effective decision because his leadership did call on him later that afternoon, and he was already in a position to start responding to their needs.

As we have already noted, most strategists are valued for their knowledge capital, and this means that most work in larger, hierarchical organizations. Most of the strategists with whom we spoke were senior people at the tops of their respective hierarchies, and this means that they relied heavily on direct reports and trusted staff members whenever a surprise occurred. Therefore, we observed that the next step strategists took after taking steps to control their emotions was to assemble the staff and develop a plan for moving forward. In doing this, all of the strategists noted that it is very important to decide on a plan, socialize that plan throughout the organization, and present a unified front when executing that plan.

> ## *We Also Observed Two Key Findings That Were Not Part of Our Initial Hypotheses*
>
> ### We were surprised to observe that
> 1. **The most chaotic environments all feature surprises that are motivated by other humans rather than the environment.**
> 2. **The largest surprises come from third parties rather than a direct stakeholder or adversary.**
>
> **RAND**

As researchers, we were surprised by two findings that were not part of our initial hypotheses. First, as we have already mentioned, the most chaotic environments all feature surprises that are driven by other humans rather than the operating environment. Furthermore, we observed that the biggest surprises tend to come from third parties rather than a direct stakeholder or adversary.

Our initial insight into this idea that surprises tend to come from third parties came from the Navy SEAL, who noted that he did not generally encounter major surprises from his adversaries. After all, he had spent a lot of time thinking about what motivated them, along with determining what calculus the adversaries were likely using to achieve their goals. Going into a mission, he would have spent time thinking about their objectives and would have a reasonably good chance of being able to predict how his adversary would behave.

So what surprises the Navy SEAL more than his adversary? Our practitioner told us that domesticated animals and civilians present the largest surprises to his team while on a foot patrol. These two groups represent big surprises because their motivations and actions cannot be easily predicted by the SEALs. Specifically, civilians are particularly risky for his team to encounter because the SEALs do not know how civilians will respond to them.

These two observations lead to our recommendations.

Our Research on Surprise Yields Possible Actions for NRO Decisionmakers

- **Finding: experience, organizational memory, level-headedness, protocols, and teamwork can all help mitigate the negative effects of surprise.**
  - **– Possible action: Promote knowledge sharing through a common AS&T library, web-based sharing tools, and one-page briefs that highlight key work.**
- **Finding: the biggest surprises to strategists often come from third-parties; high-level strategists usually are not the first person in the org to detect the surprise.**
  - **– Possible action: Focus additional resources on entities like Africa, the Arctic, and NGOs.**
  - **– Possible action: Develop a 0th-order model that identifies *additional* third parties.**

RAND

To conclude this section, we will recap our findings and present possible actions that NRO/AS&T could take to move forward.

First, we noted that experience, controlling emotions, having effective protocols, and relying on teamwork can all help mitigate the negative effects of surprise. There are several possible actions that AS&T could take based on these findings:

- First, AS&T can promote knowledge-sharing and experience-sharing through electronic means by developing a central repository that is designed to facilitate information-sharing throughout the organization. This could be achieved either through an online library or web-based tools that help employees share their experiences with one another. Recognizing that AS&T analysts are often very busy, it may be most efficient to manage knowledge-sharing by relying more on one-page briefs that concisely summarize findings rather than longer, more-involved reports.
- The second way to facilitate experience-sharing is through improved human-to-human interactions. For example, multidisciplinary core teams designed to tackle specific problems could be developed, with everyone on the team having common security credentials.

Our second key finding was that surprises tend to come from third parties, and that a high-level strategist usually is not the first person in the organization to detect the surprise. (This is often due to the fact that senior strategists are usually located near the top of an organizational hierarchy.) In practice, this finding suggests two possible actions for AS&T:

- First, the fact that the biggest surprises are most likely to come from third parties suggests that AS&T should focus some resources toward researching third-party intelligence problems. Three examples of these third parties for the IC might be Africa, the Arctic, and nongovernmental organizations (NGOs), which are playing increasing roles in world politics and third-world development. In addition, AS&T might also benefit by considering a more nuanced version of this lesson: What third parties exist in threat environments that are already established?
- Second, AS&T would likely benefit from investing in a model that identifies additional third parties that may pose a threat in the future. We make this suggestion with the knowledge that AS&T invests in developing "what if" plans against existing (known) threats. This recommendation suggests investing in models that can identify possible threats (or threat factors) in the future. After all, simply *identifying* potential threats is the first step toward mitigating any negative effects that may be caused by them.

We Hypothesize That Different Objectives Will Require Different Approaches

- **There are different ways to become more responsive:**
  - **Become more efficient (do more with less)**
  - **Increase agility (quickly adopt new technology)**
  - **Pursue the state-of-the-art (develop a brand new capability)**
- **We hypothesize that companies take different approaches depending on their <u>objectives</u>:**
  - **Pfizer reallocated resources to pursue high-risk, high-reward goals.**
  - **IBM changed their business line to avoid an existential threat.**
  - **Caterpillar reorganized to become more responsive to customer needs.**

RAND

The reason for this, we hypothesize, is that there are different ways to become more responsive. We list three ways on this slide, but there are certainly more: Some companies become more responsive by improving efficiency—they do more with less. We also note two other methods: increasing agility (by always staying current with new technologies, for example) or inventing something that is state of the art. The reason or reasons driving change will determine the methods for getting there.

For example, Pfizer decided to concentrate on anticancer and Alzheimer's drugs (Thomas, 2012; and Wilson, 2011). In order to do this, it sold off and divested all of its unrelated properties so it could concentrate on this high-risk, high-reward goal. In the process, it centralized its organization and processes to pursue a single mission.

By contrast, Caterpillar was interested in becoming more responsive to its customers' needs. To do this, it decentralized and set up fully contained Caterpillar offices around the country, each containing everything needed to run the business: product experts, sales and maintenance teams, and finance and accounting personnel. In doing this, it was able to customize its service to the local market, but this required the company to take a different approach than Pfizer.

Finally, IBM took yet a third approach: It fundamentally changed its product lines by choosing to focus more on consulting services rather than primarily delivering computing hardware.

One Future Research Objective Would Be to Match An Organization's Goals to Specific Processes

- **Why do it?**
  - **Cost effective**
  - **Reliability**
  - **Agility**
  - **State-of-the-art**
  - **Disruptive products**

- **How to do it?**
  - **In house v. outsourced**
  - **Inline R&D v. skunkworks**
  - **Singular v. multidisciplinary workforce**
  - **Specific v. diverse research portfolio**

RAND

We observed that Pfizer, Caterpillar, and IBM each took a different approach for becoming more responsive in their responsive marketplaces. The next step for this research would be to develop a formal framework that connects the methods with the specific goals. In other words: Is it possible to develop guidance that instructs the organization on the best way to achieve those goals? We present these ideas here for future researchers who may be interested in pursuing these topics.

For example, we recognize that there are several decisions that organizations have to make when determining how to become more responsive. What aspects of the organization should be changed? Should R&D be performed in house, or should it be outsourced? Should the company rely on strategic investments for new intellectual property, or should it develop its own set of highly skilled innovators?

In addition, how should the research department be organized? Should it be designed as a skunkworks, in which a few of the most capable individuals do all of the work? Or should the company democratize the process by giving every engineer a small percentage of free time to work on pet projects?

Finally, what skills should the overall workforce possess? A multidisciplinary workforce is more likely to generate a wider spread of ideas (more golden nuggets along with more rotten eggs). Alternatively, a singular workforce will, on average, have more productive ideas than the multidisciplinary force, but their golden nuggets will be less likely to be as golden as those from the multidisciplinary team.

We did not have the resources to start matching the reasons for innovating with the best mechanisms for getting there, but we describe this issue as one to be addressed in future work.

hardware. Instead, improved responsiveness comes from an investment in all three areas. New hardware requires an equally flexible staff to design and operate it.

Based on These Common Lessons, We Propose Three Topics for Future Research

- **Finding: A wide FOV can avert surprise.**
    - **Task: Develop a wide FOV model to help AS&T expand their "surprise space."**
- **Finding: Modularity can limit how disruptive a technology could be.**
    - **Task: How do engineered systems evolve in disruptive ways?**
- **Finding: Organizations innovate for different reasons.**
    - **Task: What organizational structures are best for achieving a particular goal?**

RAND

We conclude this report by proposing three topics for future research.

As we noted earlier, the research on surprise taught us that adopting a wide FOV can help mitigate the negative effects caused by surprise. Therefore, we propose developing a first-order wide FOV model to help AS&T expand its view over the surprise space. There are several methods that could be used to address this task: RAND's Delphi Method and RAND's existing work on robust decisionmaking could both be used to generate a list of the most likely third-party actors that may surprise the IC in the coming years, along with a set of scenarios that represent likely outcomes from these surprises.

Our research on modularity highlighted the fact that modular systems do pose limits on how disruptive a particular technology can be. Disruptive technologies create products or services in unexpected ways, and modularity's frozen interfaces put some restriction on how much a product is able to change in the future.

Swiffer® is a good example of a disruptive technology. Developed by Proctor and Gamble, Swiffer uses a woven dry cloth to clean floors without using any liquids. Before Swiffer, people usually cleaned their floors with a mop, a bucket, and some hot soapy water. When Swiffer was being tested in house, Proctor and Gamble's experts in cleaning floors noted that Swiffer was unlikely to succeed because it did not perform as well as conventional wet cleaning methods. This is certainly true, but what these experts failed to realize was that many consumers had never wet-mopped their floors because it was too much work to get out the mop and the bucket of water. When Swiffer was introduced to the market, it was well received by a large section of the population that had never mopped before, and Swiffer effectively opened a new market for mopping floors using dry cleaning methods.

One of the implied goals of the IC is to create and field more disruptive products. We propose performing research that looks to the biological sciences to develop a set of factors that highlight the conditions needed to generate disruptive products. For example, what environmental factors have led to some of nature's most disruptive patterns, like wings or legs? Or, on a more micro level: what conditions resulted in one species gaining more colorful plumage than a neighboring species? After examining some case studies, we propose looking to see what the engineering community can learn from these patterns. By looking at some analogs in nature, we hope to at least identify the conditions that might tend to generate disruptive products.

Finally, in our research on R&D processes, we noted that different organizations innovate for different reasons. We propose to continue this research by investigating one of the questions that we posed earlier in this briefing. Given a particular reason for innovating, what organizational measures are most likely to achieve that goal? For example, what types of problems is a skunkworks best at solving? When would it be more appropriate to conduct "inline" R&D by giving every engineer 10 percent free time to pursue pet projects, as is done at Google and 3M?

Bibliography

Alkire, Brien, John Birkler, Lisa Dolan, James Dryden, Bryce Mason, Gordon T. Lee, John F. Schank, and Michael Hayes, *Littoral Combat Ships: Relating Performance to Mission Package Inventories, Homeports, and Installation Sites,* Santa Monica, Calif.: RAND Corporation, MG-528-NAVY, 2007. As of April 25, 2013:
http://www.rand.org/pubs/monographs/MG528.html

Baiocchi, Dave, and D. Steven Fox, *Surprise! From CEOs to Navy SEALs: How Professionals Prepare for and Respond to Unexpected Events,* Santa Monica, Calif.: RAND Corporation, RR-341-NRO, forthcoming.

Baldwin, Carliss Y., telephone communication with the authors, November 19, 2012.

Baldwin, Carliss Y., and Kim B. Clark, *Design Rules Volume 1: The Power of Modularity,* Cambridge, Mass.: MIT Press, 2000.

———, "Modularity in the Design of Complex Engineering Systems," in D. Braha, A. A. Minai, and Y. Bar-Yam, eds., *Complex Engineered Systems,* Springer, 2006, pp. 175–205.

Baldwin, Carliss Y., and Joachim Henkel, "The Impact of Modularity on Intellectual Property and Value Appropriation," Harvard Business School working paper, November 20, 2012.

Bartlett, Christopher A., "Caterpillar, Inc.: George Schaefer Takes Charge," Harvard Business School Case 390-036, July 1991. Revised from original September 1989 version.

Bazerman, Max H., and Michael D. Watkins, *Predictable Surprises: The Disasters You Should Have Seen Coming, and How to Prevent Them,* Harvard Business Review Press, 2008.

de Neufville, Richard, and Stefan Scholtes, *Flexibility in Engineering Design,* Cambridge, Mass.: The MIT Press, 2011.

Deviney, N., K. Sturtevant, F. Zadeh, L. Peluso, and P. Tambor, "Becoming a Globally Integrated Enterprise: Lessons on Enabling Organizational and Cultural Change," *IBM Journal of Research and Development,* Vol., 56, No. 6, November 2012.

Duray, Rebecca, Peter T. Ward, Glenn W. Milligan, and William L. Berry, "Approaches to Mass Customization: Configurations and Empirical Evaluation," *Journal of Operations Management,* Vol. 18, 2000, pp. 605–625.

Gershenson, J. K., G. J. Prasad, and Y. Zhang, "Product Modularity: Definitions and Benefits," *Journal of Engineering Design,* Vol. 14, No. 3, 2003, pp. 295–313.

Gordon, Paul, "Caterpillar to Create New Divisions," *Peoria Journal Star,* October 9, 2008. As of March 15, 2013:
http://www.pjstar.com/business/x841515639/Caterpillar-to-create-new-divisions

Huang, Chun-Che, and Andrew Kusiak, "Modularity in Design of Products and Systems," *IEEE Transactions on Systems, Man, and Cybernetics, Part A: Systems and Humans*, Vol. 28, No. 1, January 1998, pp. 66–77.

Klein, Gary, *Sources of Power: How People Make Decisions*, MIT Press, 1998.

Lau, Antonio K. W., Richard C. M. Yam, and Esther Tang, "The Impact of Product Modularity on New Product Performance: Mediation by Product Innovativeness," *Journal of Product Innovation Management*, Vol. 28, 2011, pp. 270–284.

McCall, Morgan W., Michael M. Lombardo, and Ann M. Morrison, *Lessons of Experience: How Successful Executives Develop on the Job*, Free Press, 1998.

Neustadt, Richard E., and Ernest R. May, *Thinking in Time: The Uses of History for Decision-Makers*, Free Press, 1986.

O'Rourke, Ronald, *Navy Littoral Combat Ship (LCS) Program: Background, Issues, and Options for Congress*, Washington, D.C.: Library of Congress Congressional Research Service, 2012.

Pearlstein, Steven, "After Caterpillar's Turnaround, a Chance to Reinvent Globalization," *Washington Post*, April 19, 2006. As of March 15, 2013:
http://www.washingtonpost.com/wp-dyn/content/article/2006/04/18/AR2006041801884.html

Sanchez, Ron, "Strategic Flexibility in Product Competition," *Strategic Management Journal*, Vol. 16, No. S1, 1995, pp. 135–159.

Sanchez, Ron, and Joseph T. Mahoney, "Modularity, Flexibility, and Knowledge Management in Product and Organization Design," *Strategic Management Journal*, Vol. 17, Winter Special Issue, 1996, pp. 63–76.

Shaefer, S., "Product Design Partitions with Complementary Components," *Journal of Behavior and Organization*, Vol. 38, 1999, pp. 311–330.

Schilling, Melissa A., "Towards a General Modular Systems Theory and Its Application to Interfirm Product Modularity," *The Academy of Management Review*, Vol. 25, No. 2, 2000, pp. 312–334.

———, telephone communication with the authors, November 20, 2012.

Schilling, Melissa A., and H. Kevin Steensma, "The Use of Modular Organizational Forms: An Industry-Level Analysis," *Academy of Management Journal*, Vol. 44, No. 6, 2001, pp. 1149–1168.

Tassey, G., "Standardization in Technology-Based Markets," *Research Policy*, Vol. 29, Nos. 4–5, 2000, pp. 587–602.

Thomas, Katie, "Pfizer Races to Reinvent Itself," *New York Times*, May 1, 2012. As of March 15, 2013:
http://www.nytimes.com/2012/05/02/business/
pfizer-profit-declines-19-after-loss-of-lipitor-patent.html?_r=0

U.S. Navy personnel from Littoral Combat Ship Squadron (LCSRON) MCM, Detachment 2 (DET 2), interview, November 1, 2012.

U.S. Navy personnel from the LCS Program Element Office, interview, November 6, 2012.

Ulrich, Karl T., and Karen Tung, "Fundamentals of Product Modularity," *Issues in Design Manufacturing/Integration*, Vol. 39, 1991.

Weick, Karl E., and Kathleen M. Sutcliffe, *Managing the Unexpected: Resilient Performance in an Age of Uncertainty*, Jossey-Bass, 2007.

Weiss, Robert S., *Learning from Strangers: The Art and Method of Qualitative Interview Studies*, New York: Robert S. Weiss, 1994.

Wilson, Duff, "New Chief Revises Goals and Spending for Pfizer," *New York Times*, February 1, 2011. As of March 15, 2013:
http://www.nytimes.com/2011/02/02/health/02pfizer.html?ref=pfizerinc

Zarroli, Jim, "IBM Turns 100: The Company That Reinvented Itself," National Public Radio, June 16, 2011. As of March 15, 2013:
http://www.npr.org/2011/06/16/137203529/ibm-turns-100-the-company-that-reinvented-itself

Zhao-Hui, L., and Wang Jun-Hao, "The Analysis of Economic Efficiency of Modularity 'Design Rules,'" International Conference on E-Product E-Service and E-Entertainment (ICEEE), 2010.